Earth's Changing Surface
Surface
Weathering and Erosion

W9-BNC-701

by Kate Boehm Jerome

Table of Contents

Millmark
EDUCATION

Earth's surface changes all the time.
Water can change Earth's surface.
Talk about these photos.

Do the river rocks look rough
or smooth?

The river rocks look _____.

What do you see in the ocean
waves photo?

I see _____ and _____.

How do you think the waterfall gets
its name?

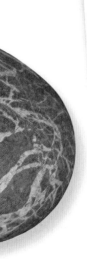

ocean waves

waterfall

rocks

water

river rocks

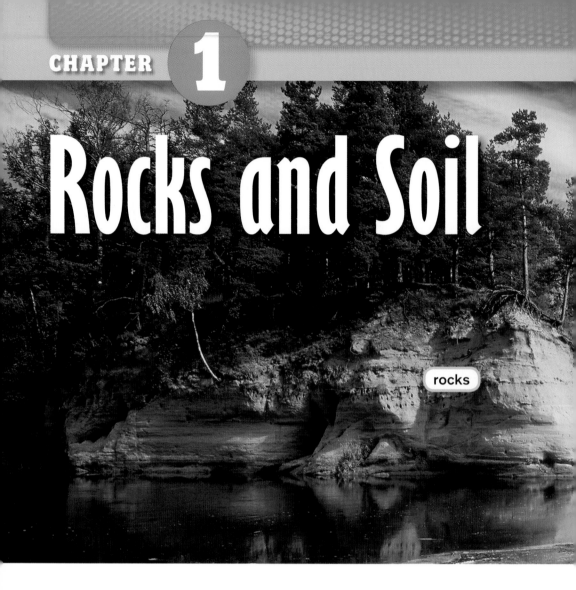

Rocks and Soil

rocks

We see rocks all over Earth's surface.
These rocks are part of Earth's **crust**.
The crust is Earth's rocky covering.

Weathering breaks
down rocks.

Most rocks in Earth's crust are hard.
But rocks can break down over time.
Weathering is how rocks break down and change.

KEY IDEA Weathering
breaks down rocks.

Weathering can break rocks into smaller pieces.
Pieces of weathered rock become part of **soil**.
Soil also has air and water in it.

When plants and animals die, their bodies break down.
This forms **humus**.
Humus becomes part of the soil, too.

KEY IDEA Soil
contains weathered
rocks, air, water,
and humus.

humus

Soil forms in layers.
Plants and their roots are usually in the top layers.
Bigger rocks are in the bottom layers.
Soil takes many years to form.

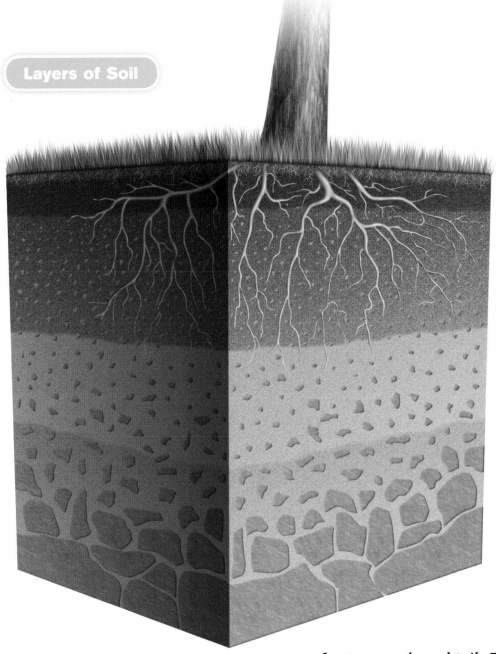

Layers of Soil

Most plants need soil to grow.
Life on Earth depends on soil and plants.

Wheat is a plant that provides
food for many living things.

KEY IDEA Most plants need soil to grow.

COMMUNICATE

Look at these photos. Talk to a friend about the photos.
Tell why you think soil is important.

MAKE CONNECTIONS

What plants and animals live in the soil near
your home? How is the soil important to you?

USE THE LANGUAGE OF SCIENCE

What are some
things that soil
contains?

Soil contains
weathered rocks,
humus, air, and water.

Chapter 1: Rocks and Soil   9

CHAPTER 2

Gravity and Wind

Weathered rocks and soil can be moved.
This movement is called **erosion**.

windblown sand and dirt

Weathering and erosion change Earth's surface.
Weathering breaks down rocks.
Erosion moves weathered rocks and soil around.

KEY IDEA Erosion moves weathered rocks and soil.

The force of wind can move sand.

Weathering and erosion are caused by **forces**.
Wind is one of these forces.

Wind can cause erosion.
Wind causes erosion when it blows sand around.

Wind can also cause weathering.
Wind blows tiny pieces of rocks and soil
against large rocks.
The tiny pieces rub against the rocks.
The rubbing slowly wears away the rocks.

▲ **Wind can wear
away rocks.**

Gravity is another force that can cause erosion.
Gravity is a force that pulls things together.
The pull of Earth's gravity is strong.
It can make loose rocks and soil slide down a hill.

KEY IDEA Wind and gravity are forces
that can cause weathering or erosion.

▼ Gravity caused
this landslide.

SUMMARIZE

This chapter is about gravity and wind.
Tell what you know about these forces.

Wind can slowly _____ large rocks.

Wind can blow tiny _____ of rock and soil
from one place to another.

Gravity pulls things _____ a hill.

Gravity can cause _____.

MAKE CONNECTIONS

Have you ever been in a
wind storm? Tell what
you saw and felt.

 ## STRATEGY FOCUS

Synthesize

Reread the ideas on page 14 and look at the photo.
Add what you already know about gravity. Make
one statement that includes most of the information.

CHAPTER 3

Water and Ice

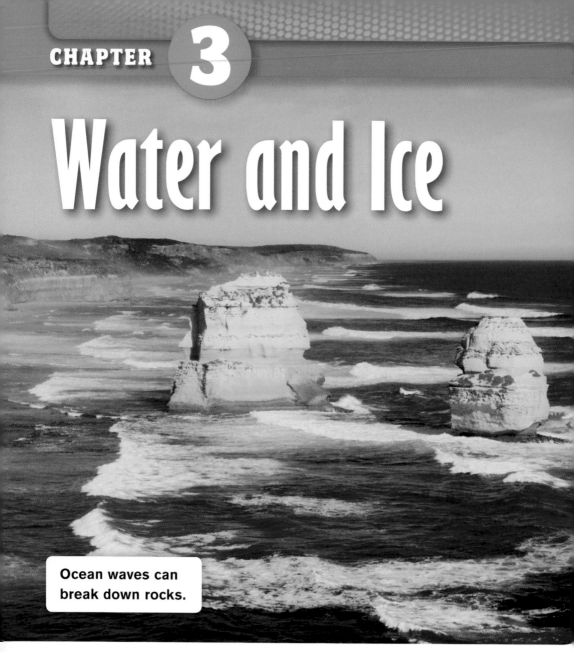

Ocean waves can break down rocks.

Water and ice can break down and move things.
They are forces that cause weathering and erosion.

The moving water in oceans hits rocks.
The force of the water changes the rocks over time.

The moving water in rivers hits rocks.
The force of the water breaks down the rocks.
The water then carries small pieces of rocks
away to other places.

Over time, this river
helped make the
Grand Canyon.

A **glacier** is a huge mass of ice.
It moves very slowly over land.

Glaciers wear away the rocks and soil under them.
Glaciers then carry the rocks and soil to other places.
Glaciers can cut deep valleys between mountains.

KEY IDEA Water and ice
are forces that can cause
weathering and erosion.

INTERPRET DATA

Country	Land Covered by Glaciers
United States	75,000 square kilometers
Canada	200,000 square kilometers
Mexico	11 square kilometers

Look at the table. Interpret data to answer the questions.

1. Which country has the most land covered by glaciers?

_____ has the most land covered by glaciers.

2. Which country has the least land covered by glaciers?

_____ has the least land covered by glaciers.

MAKE CONNECTIONS

Think of a place where you have seen water erosion. Describe the place. How has erosion changed Earth's surface?

EXPAND VOCABULARY

The word *weather* is a noun. *Weathered* can be a verb or an adjective.

Noun: The <u>weather</u> is very windy today.
Verb: The water <u>weathered</u> the rocks in the canyon.
Adjective: The <u>weathered</u> rock was shiny and smooth.

The words *force* and *layer* can also be made into nouns, verbs, or adjectives. Use these words in sentences. Use the sentences above to help.

What Is a Soil Scientist?

A soil scientist studies the upper layer of Earth's crust.

A soil scientist finds out what is in the soil.

A soil scientist helps farmers grow crops.

Would you like to be a soil scientist?

Explain your answer.

Words that Restate

When you restate something, you say the same thing with different words. You can use the word **or** to restate something. You can also use the phrase **in other words**.

EXAMPLES

A glacier moves gradually, **or** slowly, over land.

Most plants depend on soil to grow. **In other words**, plants need soil.

With a friend, look at the photos in the book. Describe a photo. Have your friend restate your description. Take turns describing and restating.

Write a Restatement

Choose three Key Ideas or captions from the book. Restate them in your own words. Use words and phrases such as **or** and **in other words**.

Words You Can Use
or
In other words
This means that

Read the newspaper article.

Have you ever seen a sinkhole?

Danger: Sinkholes!

Rocks can be worn away underground.

Sometimes a hole opens up at Earth's surface.

These are called sinkholes.

Some sinkholes are very large.

- What do you think could wear away rocks underground?

_____ could wear away rocks underground.

- Why would a big sinkhole be dangerous?

Big sinkholes could be dangerous because _____.

Key Words

crust the hard, rocky covering of Earth
The rocks you can see around you are part of Earth's **crust**.

erosion the movement of rocks and soil from one place to another
Wind causes **erosion** when it moves sand.

glacier (glaciers) a huge mass of ice
Glaciers move slowly over land.

humus a part of soil that comes from dead plants and animals
Humus is an important part of soil.

soil a material in the top layer of Earth's crust
Plants grow in **soil**.

weathering how rocks break down and change
Weathering can change Earth's surface.

Index

MILLMARK EDUCATION CORPORATION
Ericka Markman, President and CEO; Karen Peratt, VP, Editorial Director; Lisa Bingen, VP, Marketing; David Willette, VP, Sales; Rachel L. Moir, Director, Operations and Production; Shelby Alinsky, Associate Editor; Mary Ann Mortellaro, Science Editor; Amy Sarver, Series Editor; Betsy Carpenter, Editor; Guadalupe Lopez, Writer; Kris Hanneman and Pictures Unlimited, Photo Research

PROGRAM AUTHORS
Mary Hawley; Program Author, Instructional Design
Kate Boehm Jerome; Program Author, Science

BOOK DESIGN Steve Curtis Design

CONTENT REVIEWER
Tom Nolan, Operations Engineer, NASA Jet Propulsion Laboratory, Pasadena, CA

PROGRAM ADVISORS
Scott K. Baker, PhD, Pacific Institutes for Research, Eugene, OR
Carla C. Johnson, EdD, University of Toledo, Toledo, OH
Donna Ogle, EdD, National-Louis University, Chicago, IL
Betty Ansin Smallwood, PhD, Center for Applied Linguistics, Washington, DC
Gail Thompson, PhD, Claremont Graduate University, Claremont, CA
Emma Violand-Sánchez, EdD, Arlington Public Schools, Arlington, VA (retired)

TECHNOLOGY
Arleen Nakama, Project Manager
Audio CDs: Heartworks International, Inc.
CD-ROMs: Cannery Agency

PHOTO CREDITS Cover © Ross Barnett/Lonely Planet/Getty Images; 1 © Robert Fullerton/Shutterstock; 2-3 © Stock Connection Distribution/Alamy; 2a and 3b © Rosemary Calvert/Getty Images; 2b © Rolf Richardson/Alamy; 3a © Gary Crabbe/Alamy; 4-5 and 23a © Niall Benvie/Oxford Scientific Films/Photolibrary; 5 © Derrick Alderman/Alamy; 6a and 23d © Greenshoots Communications/Alamy; 6b and 23e © Kanwarjit Singh Borparai/Shutterstock; 7 illustration by Sharon and Joel Harris; 8 © Javier Larrea/age fotostock; 9a © Emilio Ereza/age fotostock; 9b © Photodisc/Punchstock; 9c © Robert W. Ginn/age fotostock; 9d © David Aubrey/Photo Researchers, Inc; 9e and 9f Lloyd Wolf for Millmark Education; 10-11 © Nigel Dennis/age fotostock; 12a and 23b © Fritz Poelking/age fotostock; 13 and 23f © Alan Majchrowicz/age fotostock; 14 © Creatas/Punchstock; 15 © Connie Cooper-Edwards/Alamy; 16 © Alex Hinds/age fotostock; 17 © Corbis/Punchstock; 18 and 23c © Jose Fuste Raga/age fotostock; 20a © Peggy Greb/USDA; 20b © Scott Bauer/USDA; 20c © Keith Weller/USDA; 21 © Joy M. Prescott/Shutterstock; 22 © AP Images/The Express-Times/Bill Adams; 24 © Henryk T. Kaiser/age fotostock

Published by Millmark Education Corporation
7272 Wisconsin Avenue, Suite 300
Bethesda, MD 20814

ISBN-13: 978-1-4334-0070-4

Printed in the USA

10 9 8 7 6 5 4 3